FREE Test Taking Tips DVD Offer

To help us better serve you, we have developed a Test Taking Tips DVD that we would like to give you for FREE. **This DVD covers world-class test taking tips that you can use to be even more successful when you are taking your test.**

All that we ask is that you email us your feedback about your study guide. Please let us know what you thought about it – whether that is good, bad or indifferent.

To get your **FREE Test Taking Tips DVD**, email freedvd@studyguideteam.com with "FREE DVD" in the subject line and the following information in the body of the email:

    a. The title of your study guide.

    b. Your product rating on a scale of 1-5, with 5 being the highest rating.

    c. Your feedback about the study guide. What did you think of it?

    d. Your full name and shipping address to send your free DVD.

If you have any questions or concerns, please don't hesitate to contact us at freedvd@studyguideteam.com.

Thanks again!

STAAR Math Grade 4

STAAR Mathematics Practice Team

Copyright © 2017 STAAR Mathematics Prep Team

All rights reserved.

Table of Contents

Quick Overview 1

Test-Taking Strategies 2

FREE DVD OFFER 6

Introduction 7

Numerical Representations and Relationships 10

    Place Value 10

    Fractions 15

Computations and Algebraic Relationships 20

    Generating Fractions 20

    Whole Number and Decimal Computations 21

    Expressions and Equations 31

Geometry and Measurement 34

    Area and Perimeter 34

    Geometric Attributes 34

    Angles 37

    Measurement 40

Data Analysis and Personal Financial Literacy 44

    Collecting, Organizing, Displaying, and Interpreting Data 44

    Managing Financial Resources 46

Practice Questions 49

Answer Explanations 57

Quick Overview

As you draw closer to taking your exam, effective preparation becomes more and more important. Thankfully, you have this study guide to help you get ready. Use this guide to help keep your studying on track and refer to it often.

This study guide contains several key sections that will help you be successful on your exam. The guide contains tips for what you should do the night before and the day of the test. Also included are test-taking tips. Knowing the right information is not always enough. Many well-prepared test takers struggle with exams. These tips will help equip you to accurately read, assess, and answer test questions.

A large part of the guide is devoted to showing you what content to expect on the exam and to helping you better understand that content. Near the end of this guide is a practice test so that you can see how well you have grasped the content. Then, answer explanations are provided so that you can understand why you missed certain questions.

Don't try to cram the night before you take your exam. This is not a wise strategy for a few reasons. First, your retention of the information will be low. Your time would be better used by reviewing information you already know rather than trying to learn a lot of new information. Second, you will likely become stressed as you try to gain a large amount of knowledge in a short amount of time. Third, you will be depriving yourself of sleep. So be sure to go to bed at a reasonable time the night before. Being well-rested helps you focus and remain calm.

Be sure to eat a substantial breakfast the morning of the exam. If you are taking the exam in the afternoon, be sure to have a good lunch as well. Being hungry is distracting and can make it difficult to focus. You have hopefully spent lots of time preparing for the exam. Don't let an empty stomach get in the way of success!

When travelling to the testing center, leave earlier than needed. That way, you have a buffer in case you experience any delays. This will help you remain calm and will keep you from missing your appointment time at the testing center.

Be sure to pace yourself during the exam. Don't try to rush through the exam. There is no need to risk performing poorly on the exam just so you can leave the testing center early. Allow yourself to use all of the allotted time if needed.

Remain positive while taking the exam even if you feel like you are performing poorly. Thinking about the content you should have mastered will not help you perform better on the exam.

Once the exam is complete, take some time to relax. Even if you feel that you need to take the exam again, you will be well served by some down time before you begin studying again. It's often easier to convince yourself to study if you know that it will come with a reward!

Test-Taking Strategies

1. Predicting the Answer

When you feel confident in your preparation for a multiple-choice test, try predicting the answer before reading the answer choices. This is especially useful on questions that test objective factual knowledge or that ask you to fill in a blank. By predicting the answer before reading the available choices, you eliminate the possibility that you will be distracted or led astray by an incorrect answer choice. You will feel more confident in your selection if you read the question, predict the answer, and then find your prediction among the answer choices. After using this strategy, be sure to still read all of the answer choices carefully and completely. If you feel unprepared, you should not attempt to predict the answers. This would be a waste of time and an opportunity for your mind to wander in the wrong direction.

2. Reading the Whole Question

Too often, test takers scan a multiple-choice question, recognize a few familiar words, and immediately jump to the answer choices. Test authors are aware of this common impatience, and they will sometimes prey upon it. For instance, a test author might subtly turn the question into a negative, or he or she might redirect the focus of the question right at the end. The only way to avoid falling into these traps is to read the entirety of the question carefully before reading the answer choices.

3. Looking for Wrong Answers

Long and complicated multiple-choice questions can be intimidating. One way to simplify a difficult multiple-choice question is to eliminate all of the answer choices that are clearly wrong. In most sets of answers, there will be at least one selection that can be dismissed right away. If the test is administered on paper, the test taker could draw a line through it to indicate that it may be ignored; otherwise, the test taker will have to perform this operation mentally or on scratch paper. In either case, once the obviously incorrect answers have been eliminated, the remaining choices may be considered. Sometimes identifying the clearly wrong answers will give the test taker some information about the correct answer. For instance, if one of the remaining answer choices is a direct opposite of one of the eliminated answer choices, it may well be the correct answer. The opposite of obviously wrong is obviously right! Of course, this is not always the case. Some answers are obviously incorrect simply because they are irrelevant to the question being asked. Still, identifying and eliminating some incorrect answer choices is a good way to simplify a multiple-choice question.

4. Don't Overanalyze

Anxious test takers often overanalyze questions. When you are nervous, your brain will often run wild, causing you to make associations and discover clues that don't actually exist. If you feel that this may be a problem for you, do whatever you can to slow down during the test. Try taking a deep breath or counting to ten. As you read and consider the question, restrict yourself to the particular words used by the author. Avoid thought tangents about what the author *really* meant, or what he or she was *trying* to say. The only things that matter on a multiple-choice test are the words that are actually in the question. You must avoid reading too much into a multiple-choice question, or supposing that the writer meant something other than what he or she wrote.

5. No Need for Panic

It is wise to learn as many strategies as possible before taking a multiple-choice test, but it is likely that you will come across a few questions for which you simply don't know the answer. In this situation, avoid panicking. Because most multiple-choice tests include dozens of questions, the relative value of a single wrong answer is small. Moreover, your failure on one question has no effect on your success elsewhere on the test. As much as possible, you should compartmentalize each question on a multiple-choice test. In other words, you should not allow your feelings about one question to affect your success on the others. When you find a question that you either don't understand or don't know how to answer, just take a deep breath and do your best. Read the entire question slowly and carefully. Try rephrasing the question a couple of different ways. Then, read all of the answer choices carefully. After eliminating obviously wrong answers, make a selection and move on to the next question.

6. Confusing Answer Choices

When working on a difficult multiple-choice question, there may be a tendency to focus on the answer choices that are the easiest to understand. Many people, whether consciously or not, gravitate to the answer choices that require the least concentration, knowledge, and memory. This is a mistake. When you come across an answer choice that is confusing, you should give it extra attention. A question might be confusing because you do not know the subject matter to which it refers. If this is the case, don't eliminate the answer before you have affirmatively settled on another. When you come across an answer choice of this type, set it aside as you look at the remaining choices. If you can confidently assert that one of the other choices is correct, you can leave the confusing answer aside. Otherwise, you will need to take a moment to try to better understand the confusing answer choice. Rephrasing is one way to tease out the sense of a confusing answer choice.

7. Your First Instinct

Many people struggle with multiple-choice tests because they overthink the questions. If you have studied sufficiently for the test, you should be prepared to trust your first instinct once you have carefully and completely read the question and all of the answer choices. There is a great deal of research suggesting that the mind can come to the correct conclusion very quickly once it has obtained all of the relevant information. At times, it may seem to you as if your intuition is working faster even than your reasoning mind. This may in fact be true. The knowledge you obtain while studying may be retrieved from your subconscious before you have a chance to work out the associations that support it. Verify your instinct by working out the reasons that it should be trusted.

8. Key Words

Many test takers struggle with multiple-choice questions because they have poor reading comprehension skills. Quickly reading and understanding a multiple-choice question requires a mixture of skill and experience. To help with this, try jotting down a few key words and phrases on a piece of scrap paper. Doing this concentrates the process of reading and forces the mind to weigh the relative importance of the question's parts. In selecting words and phrases to write down, the test taker thinks about the question more deeply and carefully. This is especially true for multiple-choice questions that are preceded by a long prompt.

9. Subtle Negatives

One of the oldest tricks in the multiple-choice test writer's book is to subtly reverse the meaning of a question with a word like *not* or *except*. If you are not paying attention to each word in the question, you can easily be led astray by this trick. For instance, a common question format is, "Which of the following is…?" Obviously, if the question instead is, "Which of the following is not…?," then the answer will be quite different. Even worse, the test makers are aware of the potential for this mistake and will include one answer choice that would be correct if the question were not negated or reversed. A test taker who misses the reversal will find what he or she believes to be a correct answer and will be so confident that he or she will fail to reread the question and discover the original error. The only way to avoid this is to practice a wide variety of multiple-choice questions and to pay close attention to each and every word.

10. Reading Every Answer Choice

It may seem obvious, but you should always read every one of the answer choices! Too many test takers fall into the habit of scanning the question and assuming that they understand the question because they recognize a few key words. From there, they pick the first answer choice that answers the question they believe they have read. Test takers who read all of the answer choices might discover that one of the latter answer choices is actually *more* correct. Moreover, reading all of the answer choices can remind you of facts related to the question that can help you arrive at the correct answer. Sometimes, a misstatement or incorrect detail in one of the latter answer choices will trigger your memory of the subject and will enable you to find the right answer. Failing to read all of the answer choices is like not reading all of the items on a restaurant menu: you might miss out on the perfect choice.

11. Spot the Hedges

One of the keys to success on multiple-choice tests is paying close attention to every word. This is never more true than with words like *almost*, *most*, *some*, and *sometimes*. These words are called "hedges" because they indicate that a statement is not totally true or not true in every place and time. An absolute statement will contain no hedges, but in many subjects, like literature and history, the answers are not always straightforward or absolute. There are always exceptions to the rules in these subjects. For this reason, you should favor those multiple-choice questions that contain hedging language. The presence of qualifying words indicates that the author is taking special care with his or her words, which is certainly important when composing the right answer. After all, there are many ways to be wrong, but there is only one way to be right! For this reason, it is wise to avoid answers that are absolute when taking a multiple-choice test. An absolute answer is one that says things are either all one way or all another. They often include words like *every*, *always*, *best*, and *never*. If you are taking a multiple-choice test in a subject that doesn't lend itself to absolute answers, be on your guard if you see any of these words.

12. Long Answers

In many subject areas, the answers are not simple. As already mentioned, the right answer often requires hedges. Another common feature of the answers to a complex or subjective question are qualifying clauses, which are groups of words that subtly modify the meaning of the sentence. If the question or answer choice describes a rule to which there are exceptions or the subject matter is complicated, ambiguous, or confusing, the correct answer will require many words in order to be expressed clearly and accurately. In essence, you should not be deterred by answer choices that seem excessively long. Oftentimes, the author of the text will not be able to write the correct answer without offering some qualifications and modifications. Your job is to read the answer choices thoroughly and completely and to select the one that most accurately and precisely answers the question.

13. Restating to Understand

Sometimes, a question on a multiple-choice test is difficult not because of what it asks but because of how it is written. If this is the case, restate the question or answer choice in different words. This process serves a couple of important purposes. First, it forces you to concentrate on the core of the question. In order to rephrase the question accurately, you have to understand it well. Rephrasing the question will concentrate your mind on the key words and ideas. Second, it will present the information to your mind in a fresh way. This process may trigger your memory and render some useful scrap of information picked up while studying.

14. True Statements

Sometimes an answer choice will be true in itself, but it does not answer the question. This is one of the main reasons why it is essential to read the question carefully and completely before proceeding to the answer choices. Too often, test takers skip ahead to the answer choices and look for true statements. Having found one of these, they are content to select it without reference to the question above. Obviously, this provides an easy way for test makers to play tricks. The savvy test taker will always read the entire question before turning to the answer choices. Then, having settled on a correct answer choice, he or she will refer to the original question and ensure that the selected answer is relevant. The mistake of choosing a correct-but-irrelevant answer choice is especially common on questions related to specific pieces of objective knowledge, like historical or scientific facts. A prepared test taker will have a wealth of factual knowledge at his or her disposal, and should not be careless in its application.

15. No Patterns

One of the more dangerous ideas that circulates about multiple-choice tests is that the correct answers tend to fall into patterns. These erroneous ideas range from a belief that B and C are the most common right answers, to the idea that an unprepared test-taker should answer "A-B-A-C-A-D-A-B-A." It cannot be emphasized enough that pattern-seeking of this type is exactly the WRONG way to approach a multiple-choice test. To begin with, it is highly unlikely that the test maker will plot the correct answers according to some predetermined pattern. The questions are scrambled and delivered in a random order. Furthermore, even if the test maker was following a pattern in the assignation of correct answers, there is no reason why the test taker would know which pattern he or she was using. Any attempt to discern a pattern in the answer choices is a waste of time and a distraction from the real work of taking the test. A test taker would be much better served by extra preparation before the test than by reliance on a pattern in the answers.

FREE DVD OFFER

Don't forget that doing well on your exam includes both understanding the test content and understanding how to use what you know to do well on the test. We offer a completely FREE Test Taking Tips DVD that covers world class test taking tips that you can use to be even more successful when you are taking your test.

All that we ask is that you email us your feedback about your study guide. To get your **FREE Test Taking Tips DVD**, email freedvd@studyguideteam.com with "FREE DVD" in the subject line and the following information in the body of the email:

- The title of your study guide.
- Your product rating on a scale of 1-5, with 5 being the highest rating.
- Your feedback about the study guide. What did you think of it?
- Your full name and shipping address to send your free DVD.

Introduction

Function of the Test

The State of Texas Assessment of Academic Readiness (STAAR) Grade 4 Math examination is part of the state's program which is designed to test students on their mastery of the curriculum standards that are based on what is known as the Texas Essential Knowledge and Skills. The testing is conducted from grade three through the completion of high school in multiple subjects. The goal of the test results is to ensure that students have successfully mastered core subjects and are thus prepared to move onto the next grade level.

This applies to students attending Texas public schools that receive state education funds. Since charter schools are considered public schools, students in charter schools are also required to take the STAAR Grade 4 Math exam as well. Students who attend private schools or who are home schooled are not required to take the exam. However, if these students wish to transfer into a Texas public school at some time during their academic career, they will need to pass the appropriate STAAR exams at that time. This also applies to out-of-state students who are transferring into Texas public schools.

The exam currently has a pass rate of around seventy-three percent.

Test Administration

The STAAR Grade 4 Math exam is offered one time each school year during the state assigned testing days, and it is only offered as a paper test. Students taking this exam are allowed to take breaks to get water or to eat a snack. However, the time taken for these types of breaks is deducted from the available test taking time. Testing accommodations that may be needed by test takers for this exam are broken down into three categories:

- Accessibility Features – i.e., signing test directions for deaf students, translating test directions for English language learners, or providing highlighters to students
- Designated Supports – i.e., large print, braille, or additional test-taking time
- Designated Supports Requiring Approval – i.e., photocopying or an additional day

These types of accommodations can be requested by completing an Accommodation Request form.

There are also three alternative versions of the STAAR Grade 4 Math exam which are relevant for specific circumstances:

- STARR L – this exam is available for English language learners
- STAAR Spanish – this exam is available for students who receive their instruction in Spanish
- STAAR Alternate 2 – this exam is available for students who receive special education services as a result of their significant cognitive disabilities

Test Format

The STAAR Grade 4 Math exam is comprised of four reporting categories. The first reporting category covers the subject area of numerical representations and relationships and is made up of nine questions. Eleven questions make up the second reporting category which is concerned with the subject area of computations and algebraic relationships. The third reporting category covers the subject area of geometry and measurement and is made up of ten questions. The final four questions make up the fourth reporting category which is concerned with data analysis and personal financial literacy.

Within each reporting category on the exam, readiness standards and supporting standards are incorporated into the various questions (as outlined in the table below). Readiness standards are more general in nature. They are concerned with important concepts that are essential for a student's success in his/her current grade, and they help to prepare a student for advancement to the next grade. Supporting standards are more narrowly defined. They are concerned with skills that underlie important concepts which may be introduced in a student's current grade, but more focus will be placed on them at a future time.

The exam contains a total of thirty-four questions. Thirty-one of the questions are in multiple choice format, and the remaining three questions are in griddable format. The griddable format is used for open-ended questions, where students are asked to transfer their answers to a separate answer sheet. They write the number going left to right in a box and then fill out the corresponding bubbles that represent the digits of the number directly below.

Examinees are not allowed to bring any type of calculator to assist them when taking this exam. Students are, however, provided with sheets of graph paper and reference materials to use during the exam, which includes some basic length, volume, weight/mass, and time conversions, a paper version of a metric ruler, and the formulas for the perimeter and area of a square and a rectangle.

Reporting Category 1		
Subject Areas	**Questions**	**Comments**
Numerical Representations & Relationships	9	3 *Readiness Standards* & 10 *Supporting Standards* are incorporated into these questions
Reporting Category 2		
Subject Areas	**Questions**	**Comments**
Computations & Algebraic Relationships	11	5 *Readiness Standards* & 7 *Supporting Standards* are incorporated into these questions
Reporting Category 3		
Subject Areas	**Questions**	**Comments**
Geometry & Measurement	10	4 *Readiness Standards* & 7 *Supporting Standards* are incorporated into these questions
Reporting Category 4		
Subject Areas	**Questions**	**Comments**
Data Analysis & Personal Financial Literacy	4	1 *Readiness Standard* & 4 *Supporting Standards* are incorporated into these questions
Total	34	31 questions are in multiple choice format & 3 questions are in griddable format
4 Hours to Complete Exam		

Scoring

Students are not penalized for guessing when answering the multiple choice questions on this exam.

Students receive cut scores on the exam which translate into one of the following four levels of performance.

Masters Grade Level Performance	Scores ≥ 1596
Meets Grade Level Performance	Scores ≥ 1486
Approaches Grade Level Performance	Scores ≥ 1360
Did Not Meet Grade Level Performance	Scores < 1360

Students who fall in the masters grade level performance category should do well in the next grade without any additional lessons. The meets grade level performance designation indicates that students will most likely succeed in the next grade. Any additional skill building needed should be only for the short term and only in specific areas. Students who are in the approaches grade level performance area will probably need more academic preparedness activities to be successful in the next grade. The did not meet grade level performance category means that students do not have an appropriate level of understanding and knowledge to succeed in the next grade without comprehensive and consistent academic instruction.

Students and parents are granted access to a secure student portal where they can log in using a unique access code and the student's date of birth to view exam results.

Numerical Representations and Relationships

Place Value

Interpreting the Value of Each Place-Value Position

In accordance with the base-10 system, the value of a digit increases by a factor of ten each place it moves to the left. For example, consider the number 7. Moving the digit one place to the left (70), increases its value by a factor of 10 ($7 \times 10 = 70$). Moving the digit two places to the left (700) increases its value by a factor of 10 twice ($7 \times 10 \times 10 = 700$). Moving the digit three places to the left (7,000) increases its value by a factor of 10 three times ($7 \times 10 \times 10 \times 10 = 7,000$), and so on.

Conversely, the value of a digit decreases by a factor of ten each place it moves to the right. (Note that multiplying by $\frac{1}{10}$ is equivalent to dividing by 10). For example, consider the number 40. Moving the digit one place to the right (4) decreases its value by a factor of 10 ($40 \div 10 = 4$). Moving the digit two places to the right (0.4), decreases its value by a factor of 10 twice ($40 \div 10 \div 10 = 0.4$) or ($40 \times \frac{1}{10} \times \frac{1}{10} = 0.4$). Moving the digit three places to the right (0.04) decreases its value by a factor of 10 three times ($40 \div 10 \div 10 \div 10 = 0.04$) or ($40 \times \frac{1}{10} \times \frac{1}{10} \times \frac{1}{10} = 0.04$), and so on.

Representing Values Using Expanded Notation and Numerals

The number system that is used consists of only ten different digits or characters. However, this system is used to represent an infinite number of values. The place value system makes this infinite number of values possible. The position in which a digit is written corresponds to a given value. Starting from the decimal point (which is implied, if not physically present), each subsequent place value to the left represents a value greater than the one before it. Conversely, starting from the decimal point, each subsequent place value to the right represents a value less than the one before it.

The names for the place values to the left of the decimal point are as follows:

...	Billions	Hundred-Millions	Ten-Millions	Millions	Hundred-Thousands	Ten-Thousands	Thousands	Hundreds	Tens	Ones

*Note that this table can be extended infinitely further to the left.

The names for the place values to the right of the decimal point are as follows:

Decimal Point (.)	Tenths	Hundredths	Thousandths	Ten-Thousandths	...

*Note that this table can be extended infinitely further to the right.

When given a multi-digit number, the value of each digit depends on its place value. Consider the number 682,174.953. Referring to the chart above, it can be determined that the digit 8 is in the ten-thousands place. It is in the fifth place to the left of the decimal point. Its value is 8 ten-thousands or 80,000. The digit 5 is two places to the right of the decimal point. Therefore, the digit 5 is in the hundredths place. Its value is 5 hundredths or $\frac{5}{100}$ (equivalent to .05).

A number is written in expanded form by expressing it as the sum of the value of each of its digits. The expanded form of 2,356,471.89, which is written with the highest value first down to the lowest value, is expressed as: 2,000,000 + 300,000 + 50,000 + 6,000 + 400 + 70 + 1 + .8 + .09 7,000 + 600 + 30 + 1 + .4 + .02.

When verbally expressing a number, the integer part of the number (the numbers to the left of the decimal point) resembles the expanded form without the addition between values. In the above example, the numbers read "seven thousand six hundred thirty-one." When verbally expressing the decimal portion of a number, the number is read as a whole number, followed by the place value of the furthest digit (non-zero) to the right. In the above example, 0.42 is read "forty-two hundredths." Reading the number 7,631.42 in its entirety is expressed as "seven thousand six hundred thirty-one and forty-two hundredths." The word *and* is used between the integer and decimal parts of the number.

Comparing and Ordering Whole Numbers

Rational numbers are any number that can be written as a fraction or ratio. Within the set of rational numbers, several subsets exist that are referenced throughout the mathematics topics. Counting numbers are the first numbers learned as a child. Counting numbers consist of 1,2,3,4, and so on. Whole numbers include all counting numbers and zero (0,1,2,3,4,…). Integers include counting numbers, their opposites, and zero (…,-3,-2,-1,0,1,2,3,…). Rational numbers are inclusive of integers, fractions, and decimals that terminate, or end (1.7, 0.04213) or repeat ($0.13\overline{65}$).

Placing numbers in an order in which they are listed from smallest to largest is known as *ordering*. Ordering numbers properly can help in the comparison of different quantities of items.

When comparing two numbers to determine if they are equal or if one is greater than the other, it is best to look at the digit furthest to the left of the decimal place (or the first value of the decomposed numbers). If this first digit of each number being compared is equal in place value, then move one digit to the right to conduct a similar comparison. Continue this process until it can be determined that both numbers are equal or a difference is found, showing that one number is greater than the other. If a number is greater than the other number it is being compared to, a symbol such as > (greater than) or < (less than) can be utilized to show this comparison. It is important to remember that the "open mouth" of the symbol should be nearest the larger number.

For example:

1,023,100 compared to 1,023,000

First, compare the digit farthest to the left. Both are decomposed to 1,000,000, so this place is equal.

Next, move one place to right on both numbers being compared. This number is zero for both numbers, so move on to the next number to the right. The first number decomposes to 20,000, while the second decomposes to 20,000. These numbers are also equal, so move one more place to the right. The first number decomposes to 3,000, as does the second number, so they are equal again. Moving one place to the right, the first number decomposes to 100, while the second number is zero. Since 100 is greater than zero, the first number is greater than the second. This is expressed using the greater than symbol:

1,023,100 > 1,023,000 because 1,023,100 is greater than 1,023,000 (Note that the "open mouth" of the symbol is nearest to 1,023,100).

Another way to compare whole numbers with many digits is to use place value. In each number to be compared, it is necessary to find the highest place value in which the numbers differ and to compare the value within that place value. For example, 4,523,345 < 4,532,456 because of the values in the ten thousands place.

Rounding Whole Numbers

Rounding numbers changes the given number to a simpler and less accurate number than the exact given number. Rounding allows for easier calculations which estimate the results of using the exact given number. The accuracy of the estimate and ease of use depends on the place value to which the number is rounded. Rounding numbers consists of:

- Determining what place value the number is being rounded to
- Examining the digit to the right of the desired place value to decide whether to round up or keep the digit
- Replacing all digits to the right of the desired place value with zeros

To round 746,311 to the nearest ten thousands, the digit in the ten thousands place should be located first. In this case, this digit is 4 (7**4**6,311). Then, the digit to its right is examined. If this digit is 5 or greater, the number will be rounded up by increasing the digit in the desired place by one. If the digit to the right of the place value being rounded is 4 or less, the number will be kept the same. For the given example, the digit being examined is a 6, which means that the number will be rounded up by increasing the digit to the left by one. Therefore, the digit 4 is changed to a 5. Finally, to write the rounded number, any digits to the left of the place value being rounded remain the same and any to its right are replaced with zeros. For the given example, rounding 746,311 to the nearest ten thousand will produce 750,000. To round 746,311 to the nearest hundred, the digit to the right of the three in the hundreds place is examined to determine whether to round up or keep the same number. In this case, that digit is a one, so the number will be kept the same and any digits to its right will be replaced with zeros. The resulting rounded number is 746,300.

Rounding place values to the right of the decimal follows the same procedure, but digits being replaced by zeros can simply be dropped. To round 3.752891 to the nearest thousandth, the desired place value is located (3.75**2**891) and the digit to the right is examined. In this case, the digit 8 indicates that the number will be rounded up, and the 2 in the thousandths place will increase to a 3. Rounding up and replacing the digits to the right of the thousandths place produces 3.753000 which is equivalent to 3.753. Therefore, the zeros are not necessary and the rounded number should be written as 3.753.

When rounding up, if the digit to be increased is a 9, the digit to its left is increased by 1 and the digit in the desired place value is changed to a zero. For example, the number 1,598 rounded to the nearest ten is 1,600. Another example shows the number 43.72961 rounded to the nearest thousandth is 43.730 or 43.73.

Representing Decimals

Decimals and fractions are two ways that can be used to represent positive numbers less than one. Counting money—specifically, quantities less than one dollar—is a good method to introduce values less than one as problems involving change are important story problems that are applicable to real-world situations. For example, if a student had three quarters and a dime and wanted to purchase a cookie at lunch for 50 cents, how much change would she receive? The answer is found by first calculating the sum of the change as 85 cents and then subtracting 50 cents to get 35 cents. Money can also be utilized as a technique to learn the transition and relationship between decimals and fractions. For example, a dime represents $0.10 or $1/10$ of a dollar. Problems involving both dollars and cents also can be introduced. For example, if someone has three dollar bills and two quarters, the amount can be represented as a decimal as $3.50.

Formally, a *decimal* is a number that has a dot within the number. For example, 3.45 is a decimal, and the dot is called a *decimal point*. The number to the left of the decimal point is in the ones place. The number to the right of the decimal point represents the portion of the number less than one. The first number to the right of the decimal point is the tenths place, and one tenth represents $1/10$, just like a dime. The next place is the hundredths place, and it represents $1/100$, just like a penny. This idea is continued to the right in the hundredths, thousandths, and ten thousandths places. Each place value to the right is ten times smaller than the one to its left.

A number less than one contains only digits in some decimal places. For example, 0.53 is less than one. A *mixed number* is a number greater than one that also contains digits in some decimal places. For example, 3.43 is a mixed number. Adding a zero to the right of a decimal does not change the value of the number. For example, 2.75 is the same as 2.750. However, 2.75 is the more accepted representation of the number. Also, zeros are usually placed in the ones column in any value less than one. For example, 0.65 is the same as .65, but 0.65 is more widely used.

In order to read or write a decimal, the decimal point is ignored. The number is read as a whole number, and then the place value unit is stated in which the last digit falls. For example, 0.089 is read as *eighty-nine thousandths*, and 0.1345 is read as *one thousand, three hundred forty-five ten thousandths*. In mixed numbers, the word "and" is used to represent the decimal point. For example, 2.56 is read as *two and fifty-six hundredths*.

Comparing and Ordering Decimals

To compare decimals and order them by their value, utilize a method similar to that of ordering large numbers.

The main difference is where the comparison will start. Assuming that any numbers to left of the decimal point are equal, the next numbers to be compared are those immediately to the right of the decimal point. If those are equal, then move on to compare the values in the next decimal place to the right.

For example:

Which number is greater, 12.35 or 12.38?

Check that the values to the left of the decimal point are equal:

$$12 = 12$$

Next, compare the values of the decimal place to the right of the decimal:

$$12.3 = 12.3$$

Those are also equal in value.

Finally, compare the value of the numbers in the next decimal place to the right on both numbers:

$$12.3\mathbf{5} \text{ and } 12.3\mathbf{8}$$

Here the 5 is less than the 8, so the final way to express this inequality is:

12.35 < 12.38

Comparing decimals is regularly exemplified with money because the "cents" portion of money ends in the hundredths place. When paying for gasoline or meals in restaurants, and even in bank accounts, if enough errors are made when calculating numbers to the hundredths place, they can add up to dollars and larger amounts of money over time.

Number lines can also be used to compare decimals. Tick marks can be placed within two whole numbers on the number line that represent tenths, hundredths, etc. Each number being compared can then be plotted. The leftmost value on the number line is the largest.

Determining Decimals on a Number Line

To precisely understand a number being represented on a number line, the first step is to identify how the number line is divided up. When utilizing a number line to represent decimal portions of numbers, it is helpful to label the divisions, or insert additional divisions, as needed.

For example, what number, to the nearest hundredths place is marked by the point on the following number line?

First, figure out how the number line is divided up. In this case, it has ten sections, so it is divided into tenths. To use this number line with the divisions, label the divisions as follows:

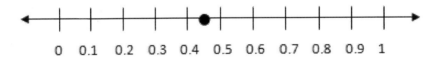

Because the dot is placed equally between 0.4 and 0.5, it is at 0.45.

What number, to the nearest tenths place, is marked by the point on the following number line?

First, determine what the number line is divided up into and mark it on the line.

This number line is divided up into half of the whole numbers it represents, or 0.5 increments. The division and labeling of the number line assists in easily reading the dot as marking 2.5.

Fractions

Representing Fractions

A *fraction* is a part of something that is whole. Items such as apples can be cut into parts to help visualize fractions. If an apple is cut into 2 equal parts, each part represents ½ of the apple. If each half is cut into two parts, the apple now is cut into quarters. Each piece now represents ¼ of the apple. In this example, each part is equal because they all have the same size. Geometric shapes, such as circles and squares, can also be utilized in the classroom to help visualize the idea of fractions.

For example, a circle can be drawn on the board and divided into 6 equal parts:

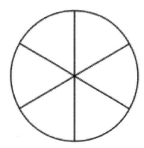

Shading can be used to represent parts of the circle that can be translated into fractions. The top of the fraction, the *numerator*, can represent how many segments are shaded. The bottom of the fraction, the *denominator*, can represent the number of segments that the circle is broken into. A pie is a good analogy to use in this example. If one piece of the circle is shaded, or one piece of pie is cut out, ¹/₆ of the object is being referred to. An apple, a pie, or a circle can be utilized in order to compare simple fractions. For example, showing that ¹/₂ is larger than ¹/₄ and that ¹/₄ is smaller than ¹/₃ can be accomplished through shading. A *unit fraction* is a fraction in which the numerator is 1, and the denominator is a positive whole number. It represents one part of a whole—one piece of pie.

When representing fractions, it is important to remember the meaning of the word *fraction*. The number is a fraction, or portion, of a whole. The whole, or 1, is the number on the bottom part of the fraction, the denominator. For example, the fraction $\frac{2}{5}$ represents a number of portions (the numerator, 2) of how many it would take to make a whole (the denominator, 5). So, $\frac{2}{5}$ represents 2 portions out of the 5 it would take to make a whole.

This could also be represented with blocks, as follows:

5 blocks = 1 whole, so only 2 of the 5 are shaded.

This method could also be used to represent fractions with a higher number in the numerator than in the denominator.

What does the fraction $\frac{6}{5}$ look like with the block method?

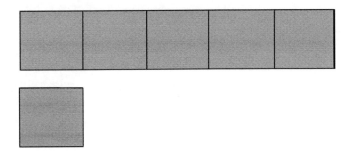

Decomposing Fractions

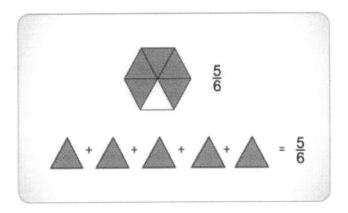

Fractions can be broken apart into sums of fractions with the same denominator. For example, the fraction $\frac{5}{6}$ can be decomposed into sums of fractions with all denominators equal to 6 and the numerators adding to 5. The fraction $\frac{5}{6}$ is decomposed as an of the following:

$$\frac{3}{6}+\frac{2}{6}$$

$$\frac{2}{6}+\frac{2}{6}+\frac{1}{6}$$

$$\frac{3}{6}+\frac{1}{6}+\frac{1}{6}$$

$$\frac{1}{6}+\frac{1}{6}+\frac{1}{6}+\frac{2}{6}$$

$$\frac{1}{6}+\frac{1}{6}+\frac{1}{6}+\frac{1}{6}+\frac{1}{6}$$

A unit fraction is a fraction in which the numerator is 1. If decomposing a fraction into unit fractions, the sum will consist of a unit fraction added the number of times equal to the numerator. For example, $\frac{3}{4} = \frac{1}{4}+\frac{1}{4}+\frac{1}{4}$ (unit fractions $\frac{1}{4}$ added 3 times). Composing fractions is simply the opposite of decomposing. It is the process of adding fractions with the same denominators to produce a single fraction. For example, $\frac{3}{7} + \frac{2}{7} = \frac{5}{7}$ and $\frac{1}{5} + \frac{1}{5} + \frac{1}{5} = \frac{3}{5}$.

Identifying Equivalent Fractions

Like fractions, or *equivalent fractions*, represent two fractions that are made up of different numbers, but represent the same quantity. For example, the given fractions are ⁴/₈ and ³/₆. If a pie was cut into 8 pieces and 4 pieces were removed, half of the pie would remain. Also, if a pie was split into 6 pieces and 3 pieces were eaten, half of the pie would also remain. Therefore, both of the fractions represent half of a pie. These two fractions are referred to as like fractions. *Unlike fractions* are fractions that are different and cannot be thought of as representing equal quantities. When working with fractions in mathematical expressions, like fractions should be simplified. Both ⁴/₈ and ³/₆ can be simplified into ¹/₂.

Comparing fractions can be completed through the use of a number line. For example, if $\frac{3}{5}$ and $\frac{6}{10}$ need to be compared, each fraction should be plotted on a number line. To plot $\frac{3}{5}$, the area from 0 to 1 should be broken into 5 equal segments, and the fraction represents 3 of them. To plot $\frac{4}{10}$, the area from 0 to 1 should be broken into 10 equal segments and the fraction represents 6 of them.

It can be seen that $\frac{3}{5} = \frac{6}{10}$

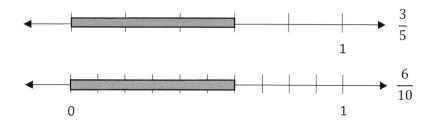

Like fractions are plotted at the same point on a number line.

Comparing Fractions

Comparing two fractions with different denominators can be difficult if attempting to guess at how much each represents. Using a number line, blocks, or just finding a common denominator with which to compare the two fractions makes this task easier.

For example, compare the fractions $\frac{3}{4}$ and $\frac{5}{8}$.

The number line method of comparison involves splitting one number line evenly into 4 sections, and the second number line evenly into 8 sections total, as follows:

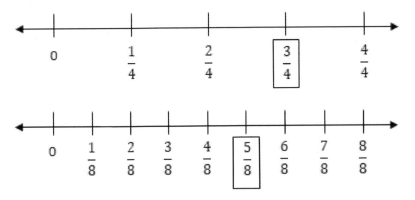

Here it can be observed that $\frac{3}{4}$ is greater than $\frac{5}{8}$, so the comparison is written as $\frac{3}{4} > \frac{5}{8}$.

This could also be shown by finding a common denominator for both fractions, so that they could be compared. First, list out factors of 4: 4, 8, 12, 16.

Then, list out factors of 8: 8, 16, 24.

Both share a common factor of 8, so they can be written in terms of 8 portions. In order for $\frac{3}{4}$ to be written in terms of 8, both the numerator and denominator must be multiplied by 2, thus forming the new fraction $\frac{6}{8}$. Now the two fractions can be compared.

Because both have the same denominator, the numerator will show the comparison.

$$\frac{6}{8} > \frac{5}{8}$$

Representing Fractions and Decimals on a Number Line

To represent fractions and decimals as distances beginning at zero on a number line, it's helpful to relate the fraction to a real-world application. For example, a charity walk covers $\frac{3}{10}$ of a mile. How could this distance be represented on a number line?

First, divide the number line into tenths, as follows:

If each division on the number line represents one-tenth of one, or $\frac{1}{10}$, then representing the distance of the charity walk, $\frac{3}{10}$, would cover 3 of those divisions and look as follows:

So, the fraction $\frac{3}{10}$ is represented by covering from 0 to 0.3 (or 3 sections) on the number line.

Computations and Algebraic Relationships

Generating Fractions

Solving Addition and Subtraction of Fraction

When adding or subtracting fractions, the numbers must have the same denominators. In these cases, numerators are added or subtracted and denominators are kept the same. For example, $\frac{2}{7} + \frac{3}{7} = \frac{5}{7}$ and $\frac{4}{5} - \frac{3}{5} = \frac{1}{5}$.

A common mistake would be to add the denominators so that

$$\frac{1}{4} + \frac{1}{4} = \frac{1}{8} \text{ or } \frac{2}{8}$$

However, conceptually, it is known that two quarters make a half, so neither one of these are correct.

If fractions have the same denominator, the numerators of the fractions can be combined using addition and subtraction methods.

For example, what would $\frac{3}{5} + \frac{1}{5}$ equal?

Adding the numerators and then placing the solution over the original denominator works, as follows:

$$3 + 1 = 4$$

Therefore, $\frac{3}{5} + \frac{1}{5} = \frac{4}{5}$.

Another way to show this is on a number line:

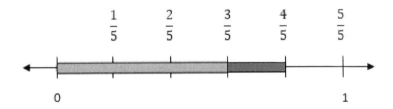

This could also be done by showing jumps on the number line for each division:

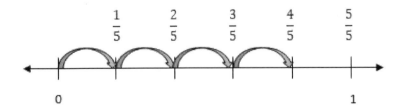

Either method also works for demonstrating subtraction. For example, what is $\frac{4}{5} - \frac{2}{5}$?

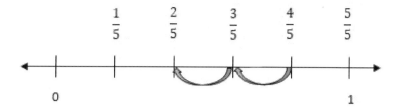

Using the jumps method on the number line, start at $\frac{4}{5}$ and subtract 2 divisions.

The end position is at the $\frac{2}{5}$ mark.

Because both fractions have the same denominator, the result of subtracting the numerators is as follows:

$$4 - 2 = 2$$

Then place this number over the original common denominator of 5 for the solution of $\frac{2}{5}$.

Evaluate Sums and Differences of Fractions

To evaluate if a sum or difference seems reasonable, knowing certain benchmark fractions is useful. That way, it is easy to gauge whether a solution is probable or not. For example, add the fractions $\frac{1}{5} + \frac{1}{3}$.

To calculate this solution, a common denominator must be established, and the fractions must be rewritten with their new numerators and common denominator.

$$\frac{1}{5} + \frac{1}{3} = \left(\frac{1}{5} \times \frac{3}{3}\right) + \left(\frac{1}{3} \times \frac{5}{5}\right)$$

$$= \frac{3}{15} + \frac{5}{15} = \frac{8}{15}$$

When considering the original fractions, $\frac{1}{5}$ is greater than the benchmark of 0, but less than the benchmark fraction of $\frac{1}{4}$. Additionally, $\frac{1}{3}$ is more than the benchmark fraction of $\frac{1}{4}$, but less than the benchmark fraction of $\frac{1}{2}$. So, it is reasonable to estimate that the combination of these two fractions would result in a number larger than $\frac{1}{4}$, and slightly greater than $\frac{1}{2}$. The solution of $\frac{8}{15}$ is slightly more than $\frac{1}{2}$, which is a reasonable solution.

Whole Number and Decimal Computations

Adding and Subtracting Numbers

Adding and subtracting numbers with more than one digit can be done using place value and rewriting numbers in expanded form. In the addition problem $256 + 261$, 256 can be thought of as $200 + 50 + 6$ or 2 hundreds, 5 tens, and 6 ones. 261 can be thought of as $200 + 60 + 1$ or 2 hundreds, 6 tens, and 1 one. Adding the two numbers by place value results in 4 hundreds, 11 tens, and 7 ones. The 11 tens need to be regrouped as 1 hundred and 1 one. This leaves 5 hundreds, 1 ten, and 7 one, which is 517.

One method of subtraction involves a counting-up procedure. In the subtraction problem $476 - 241$, adding 9 to 241 gives 250, adding 26 to 150 gives 276, and adding 200 to 276 gives 476. Therefore, the answer to the subtraction problem is $9 + 26 + 200 = 235$. The answer can be checked by adding $235 + 241$ to make sure it equals 476. Also, the place value technique used within addition can be used rewriting each number in expanded form and then subtracting within each place value. Therefore, $400 + 70 + 6 - (200 + 40 + 1) = 200 + 30 + 5 = 235$. If one of the subtraction problems is not possible within a place value, the next largest place value must be regrouped. For instance, $262 - 71 = 200 + 60 + 2 - (70 + 1) = 100 + 160 + 2 - (70 - 2) = 100 + 90 + 1 = 191$.

When adding and subtracting numbers with decimals, the decimals need to be lined up. Zeros need to be added onto the right of any number in the decimal places to ensure the same number of decimal places in both addends. Then, addition is performed in the same manner as with whole numbers, making sure to input the decimal point into the correct place in the answer. For example, $3.5 + 2.75 = 3.50 + 2.75 = 6.25$.

To subtract numbers with decimal places, the key is to line up the decimal points and begin normal subtraction.

For example, subtract $15.67 - 2.13$.

First, line up the decimal points, then subtract, as usual:

```
   15.67
-   2.13
   13.54
```

Notice the answer maintains the position of the decimal point.

If two numbers do not have the same number of decimal places, zeroes can be filled in as place holders.

For example, subtract $9.54 - 2.2$.

First, line up the decimal points.

```
   9.54
-  2.2
```

Next, fill in any zeroes necessary for place holders (the number 2.2 needs one more decimal place to match the number in 9.54). Then, subtract as usual.

```
   9.54
-  2.20
   7.34
```

The solution maintains the decimal point in the correct position.

Determining Products Using Properties of Operations and Place Value

Multiplication can be completed using place value. When a number is multiplied times 10, the number shifts over one place value to the left, and a 0 is entered in the ones place. For example, $1{,}235 \times 10 = 12{,}350$. Similarly, when a number is multiplied times 100, the entire number is shifted over two place values to the left, and a 0 is entered in both the ones and tens places. For example, $15{,}634 \times 100 = 1{,}563{,}400$. This same technique can be used to multiply single digit numbers times factors of 10 and 100. For instance, 5×300 can be thought of as $5 \times 3 \times 100 = 15 \times 100 = 1{,}500$.

Multiplication also adheres to certain properties of operations. When reviewing calculations consisting of more than one operation, the order in which the operations are performed affects the resulting answer. Consider $5 \times 2 + 7$. Performing multiplication then addition results in an answer of 17 ($5 \times 2 = 10$; $10 + 7 = 17$). However, if the problem is written $5 \times (2 + 7)$, the order of operations dictates that the operation inside the parenthesis must be performed first. The resulting answer is 45 ($2 + 7 = 9$, then $5 \times 9 = 45$).

The order in which operations should be performed is remembered using the acronym PEMDAS. PEMDAS stands for parenthesis, exponents, multiplication/division, and addition/subtraction. Multiplication and division are performed in the same step, working from left to right with whichever comes first. Addition and subtraction are performed in the same step, working from left to right with whichever comes first.

Consider the following example: $8 \div 4 + 8(7 - 7)$. Performing the operation inside the parenthesis produces $8 \div 4 + 8(0)$ or $8 \div 4 + 8 \times 0$. There are no exponents, so multiplication and division are performed next from left to right resulting in: $2 + 8 \times 0$, then $2 + 0$. Finally, addition and subtraction are performed to obtain an answer of 2. Now consider the following example: $6x3 + 3^2 - 6$. Parentheses are not applicable. Exponents are evaluated first, $6 \times 3 + 9 - 6$. Then multiplication/division forms $18 + 9 - 6$. At last, addition/subtraction leads to the final answer of 21.

With any number times one (for example, $8 \times 1 = 8$) the original amount does not change. Therefore, one is the *multiplicative identity*. For any whole number a, $1 \times a = a$. Also, any number multiplied times zero results in zero. Therefore, for any whole number a, $0 \times a = 0$.

Multiplication also follows the commutative property. The order in which multiplication is calculated does not matter. For example, 3 x 10 and 10 x 3 both equal 30. Ten sets of three apples and three sets of ten apples both equal 30 apples. The *commutative property of multiplication* states that for any whole numbers a and b, $a \times b = b \times a$. Multiplication also follows the associative property because the product of three or more whole numbers is the same, no matter what order the multiplication is completed. The *associative property of multiplication* states that for any whole numbers a, b, and c, $(a \times b) \times c = a \times (b \times c)$. For example, $(2 \times 3) \times 4 = 2 \times (3 \times 4)$.

The *distributive property of multiplication over addition* is an extremely important concept that appears in algebra. It states that for any whole numbers a, b, and c, it is true that $a \times (b + c) = (a \times b) + (a \times c)$. Because multiplication is commutative, it is also true that $(b + c) \times a = (b \times a) + (c \times a)$. For example, $100 \times (3 + 2)$ is the same as $(100 \times 3) + (100 \times 2)$. Both result in 500.

Representing the Product of 2 Two-Digit Numbers

A simple way to represent the product of 2 two-digit numbers is through the use of arrays.

Consider this example of the product of 12 × 12 represented by an array:

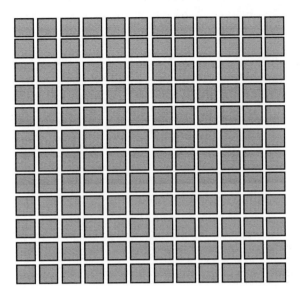

Since the resulting shape is square (12 tiles on each side) the total number of tiles represents a number that is called a *perfect square*, which is the product of a number multiplied by itself. An equation to describe this situation is 12 × 12 = 144.

144 is a perfect square.

Numbers that are not perfect squares can be represented in a similar fashion. For example, the product of 11 × 10 represented in an array looks like this:

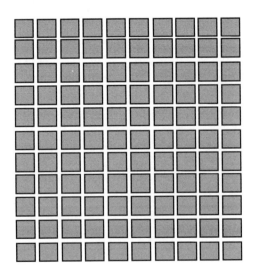

An equation to describe this situation is $11 \times 10 = 110$. This is not a square, so the total number of tiles is not a perfect square.

Any numbers can be represented using arrays, and then by an equation that describes the size of the array.

Another method of multiplication can be done with the use of an *area model*. An area model is a rectangle that is divided into rows and columns that match up to the number of place values within each number. For example, $29 \times 65 = 25 + 4$ and $66 = 60 + 5$. The products of those 4 numbers are found within the rectangle and then summed up to get the answer. The entire process is: $(60 \times 25) + (5 \times 25) + (60 \times 4) + (5 \times 4) = 1,500 + 240 + 125 + 20 = 1,885$. Here is the actual area model:

	25	4
60	60x25 1,500	60x4 240
5	5x25 125	5x4 20

```
  1,500
    240
    125
+    20
-------
  1,885
```

Using Strategies to Multiply Up to a Four-Digit Number by a One-Digit Number

One of the quickest methods to multiply larger numbers involves an *algorithm* to line up the products.

For example, multiply $1,321 \times 3$.

First, line up the numbers on the far right:

```
  1,321
×     3
```

Next, beginning on the far right and then moving one place at a time to the left, multiply the two numbers and write the answer below the line in the column of the top number. Continue the process until there are no more numbers to the left on top, as follows:

```
  1,321
×     3
      3
```

```
  1,321
×     3
     63
```

```
  1,321
×     3
    963
```

```
  1,321
×     3
  3,963
```

Some numbers will require a leading number to be carried up to the row to the left, as follows:

```
  1,621
×     3
      3
```

```
  1,621
×     3
     63
```

```
  1¹,621
×      3
     863
```

```
  1¹,621
×      3
   4,863
```

Notice that any carryovers are added to the sum of the two numbers being multiplied.

To use this algorithm for larger numbers, say a two-digit number being multiplied by another two-digit number, an additional step is necessary.

For example, multiply 12 × 15.

Line up the numbers, as before.

```
   12
× 15
```

Then begin multiplying on the far right, and move to the left.

```
  1¹2
× 1 5
    0
```

```
  1¹2
× 1 5
  6 0
```

Next, move one number to the left on the bottom row and begin the process again, while writing the products in the line under the first set of solutions.

The first row on the far right will automatically get a zero as a place holder, so the numbers will shift one column to the left of the original line up, as follows:

```
  1 2
× 1 5
  6 0
  2 0
```

```
  1 2
× 1 5
  6 0
1 2 0
```

```
  1 2
× 1 5
  6 0
1 2 0
```

Now, add the two sets of solutions.

```
  1 2
× 1 5
  6 0
1 2 0
1 8 0
```

Representing the Quotient of Up to a Four-Digit Number Divided by a One-Digit Number

Dividing a number by a single digit or two digits can be turned into repeated subtraction problems. An area model can be used throughout the problem that represents multiples of the divisor. For example, the answer to $8580 \div 55$ can be found by subtracting 55 from 8580 one at a time and counting the total number of subtractions necessary.

However, a simpler process involves using larger multiples of 55. First, $100 \times 55 = 5,500$ is subtracted from 8,580, and 3,080 is leftover. Next, $50 \times 55 = 2,750$ is subtracted from 3,080 to obtain 380. $5 \times 55 = 275$ is subtracted from 330 to obtain 55, and finally, $1 \times 55 = 55$ is subtracted from 55 to obtain zero. Therefore, there is no remainder, and the answer is $100 + 50 + 5 + 1 = 156$.

Here is a picture of the area model and the repeated subtraction process:

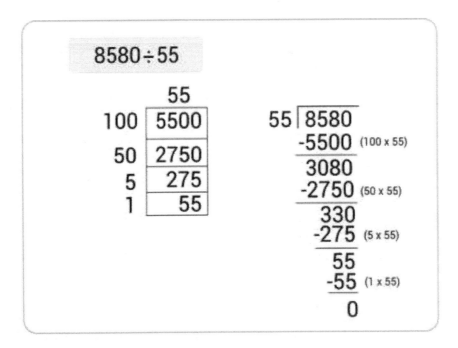

When given the task of trying to represent the quotient of up to a four-digit number divided by a one-digit number, another manageable method is utilizing an array.

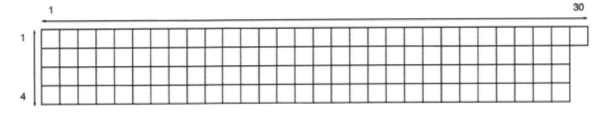

This array can be divided into three main sections. The first section is a 4 x 25 section that contains 100 pieces ($4 \times 25 = 100$). The second section is a 4 x 4 section that contains 16 pieces ($4 \times 4 = 16$). The final section is a single piece. These sections represent the division of the number 117 by 4. The sections are divided up into 4 portions of 25 plus 4 portions of 4, with one remaining portion. This leads to the following equation:

$$117 \div 4 = 25 + 4\, R1$$

$$117 \div 4 = 29\, R1$$

This shows how many times 4 can be divided into 117 and with any remainder (denoted by *R*).

Using Strategies to Divide Up to a Four-Digit Dividend by a One-Digit Divisor

To calculate a division problem, a methodical algorithm can be followed, as modeled when calculating multiplication. For each portion, the number of times a divisor can evenly go into a dividend is tracked and collected to form the final solution, or *quotient*. The process begins where the dividend and the divisor meet on the left and progresses one spot to the left after any remainder is subtracted.

For example: 375 ÷ 4.

```
    
  4|375      Set up the problem.


    9
  4|375      Because 4 cannot divide into 3, add the next unit from the numerator, 7.
   36        4 divides into 37, 9 times, so write the 9 above the 7.
   ──        Write the 36 under the 37 for subtraction; the remainder is 1 (1 is less than 4).
    1


   93
  4|375
   36↓       Drop down the next unit of the numerator, 5.
   ──        4 divides into 15, 3 times, so write the 3 above the 5.
   15        Multiply 4 x 3.
   12        Write the 12 under the 15 for subtraction; the remainder is 3 (3 is less than 4).
   ──
    3
```

The solution is 93 remainder 3 OR $93\frac{3}{4}$ (the remainder can be written over the original denominator).

Rounding to the Nearest 10, 100, or 1000

Estimation is an important tool in mathematics and in science. Understanding how to round numbers and compare and combine them is necessary to be able to properly estimate. For example, Alex's server holds 125,678 movies, while Mim's server holds 332,102 movies. If Alex and Mim have an overlap of 52,032 movies, approximately how many unique movies would the two have if they combined collections?

First, compare the numbers and decide which place value would need to be rounded in order to make an estimate for this problem. Because the smallest number is 52,032, rounding to the ten-thousands place is necessary for proper estimation. Traditional rounding uses one place smaller than the one being rounded to, to determine if the actual place value will be rounded up or remain the same. If the number is 5 or greater, the digit in question will increase by one. If the number is less than 5, the digit in question will remain the same.

In the case of 52,032:

Look at the thousands place:

5<u>2</u>,032   Since 2 is less than 5, the 5 in the ten-thousands place will remain the same.

All of the numbers after the ten-thousands place now become zeroes:

50,000

To round the other two numbers to the appropriate place, the rounding will look like the following:

12<u>5</u>,678   Since 5 is greater than or equal to 5, the value in the ten-thousands place increases by 1.

130,000

33<u>2</u>,102   Since 2 is less than 5, the value in the ten-thousands place remains the same.

330,000

An estimate would take the following calculation:

$330{,}000 + 130{,}000 - 50{,}000 = 410{,}000$

There are approximately 410,000 unique movies in the combined collection.

Solving One- and Two-Step Problems Involving Multiplication and Division

Calculations relating to real-world expenses rarely have whole number solutions. It is good practice to be able to make one-step calculations that model real-world situations. For example, Amber spends $54 on pet food in a month. If there are no increases in price for the next 11 months, how much will Amber spend on pet food during those 11 months?

Set up the multiplication problem to see how much $54 of food times 11 is:

```
   54
 × 11
   54
 +540
  594
```

Amber would spend $594 on pet food over the next 11 months.

As another example, what if Amber had $650; how many months would this last her for pet food expenses if they were $54 per month?

Set the problem up as a division problem to see how many times 54 could divide into 650. The answer will give the number of complete months the expenses could be covered.

$$\begin{array}{r} 12 \\ 54\overline{\smash{)}650} \\ -\,54 \\ \hline 110 \\ -\,108 \\ \hline 2 \end{array}$$

The remainder does not represent a full month of expenses. Therefore, Amber's $650 would last for a full 12 months of pet food expenses.

Expressions and Equations

Representing Multi-Step Problems with an Unknown Quantity

In solving multi-step problems, the first step is to line up the available information. Then, try to decide what information the problem is asking to be found. Once this is determined, construct a strip diagram to display the known information along with any information to be calculated. Finally, the missing information can be represented by a *variable* (a letter from the alphabet that represents a number) in a mathematical equation that the student can solve.

For example, Delilah collects stickers and her friends gave her some stickers to add to her current collection. Joe gave her 45 stickers, and Aimee gave her 2 times the number of stickers that Joe gave Delilah. How many stickers did Delilah have to start with, if after her friends gave her more stickers, she had a total of 187 stickers?

In order to solve this, the given information must first be sorted out. Joe gives Delilah 45 stickers, Aimee gives Delilah 2 times the number Joe gives (2 × 45), and the end total of stickers is 187.

A strip diagram represents these numbers as follows:

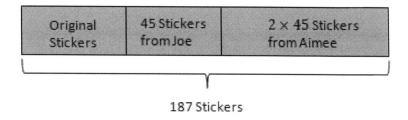

The entire situation can be modeled by this equation, using the variable s to stand for the original number of stickers:

$$s + 45 + (2 \times 45) = 187.$$

Solving for s would give the solution, as follows:

$$s + 45 + 90 = 187$$

$$s + 135 = 187$$

$$s + 135 - 135 = 187 - 135$$

$$s = 52 \text{ stickers.}$$

Using an Input-Output Table

Patterns are an important part of mathematics. Identifying and understanding how a group or pattern is represented in a problem is essential for being able to expand this process to more complex problems. A simple input-output table can model a pattern that pertains to a specific situation or equation. These can then be utilized in other areas in math, such as graphing.

For example, for every 1 parakeet the pet store sells, it sells 5 goldfish. Using the following equation to model this situation, fill in numbers missing in the input-output table, to show the total number of pets sold by the store.

Equation:

Total number of pets sold (t) = number of parakeets (p) + number of parakeets (p) × 5 goldfish

$$t = p + (p \times 5)$$
$$t = 6p$$

p	t
1	6
2	12
3	
4	24
5	

The missing numbers are 18 and 30.

This can also be shown by using an equation. If 3 is put in for p, it would look as follows:

$$t = 6 \times 3$$

$$t = 18$$

If 5 is put in for p, it would look as follows:

$$t = 6 \times 5$$

$$t = 30$$

The completed table would appear as follows:

p	t
1	6
2	12
3	18
4	24
5	30

Geometry and Measurement

Area and Perimeter

Solve Problems Related to Perimeter and Area

Perimeter is the length of all its sides. The perimeter of a given closed sided figure would be found by first measuring the length of each side and then calculating the sum of all sides. A rectangle consists of two sides called the length (*l*), which have equal measures, and two sides called the width (*w*), which have equal measures. Therefore, the perimeter (*P*) of a rectangle can be expressed as $P = l + l + w + w$. This can be simplified to produce the following formula to find the perimeter of a rectangle: $P = 2l + 2w$ or $P = 2(l + w)$.

The area of a two-dimensional figure refers to the number of square units needed to cover the interior region of the figure. This concept is similar to wallpaper covering the flat surface of a wall. For example, if a rectangle has an area of 10 square centimeters (written $10cm^2$), it will take 10 squares, each with sides one centimeter in length, to cover the interior region of the rectangle. Note that area is measured in square units such as: square centimeters or cm^2; square feet or ft^2; square yards or yd^2; square miles or mi^2.

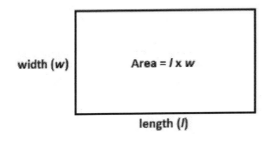

Geometric Attributes

Identifying Points, Lines, Line Segments, Rays, and Angles

The basic unit of geometry is a point. A point represents an exact location on a plane, or flat surface. The position of a point is indicated with a dot and usually named with a single uppercase letter, such as point *A* or point *T*. A point is a place, not a thing, and therefore has no dimensions or size. A set of points that lies on the same line is called collinear. A set of points that lies on the same plane is called coplanar.

The image above displays point *A*, point *B*, and point *C*.

A line is as series of points that extends in both directions without ending. It consists of an infinite number of points and is drawn with arrows on both ends to indicate it extends infinitely. Lines can be named by two points on the line or with a single, cursive, lower case letter. The two lines below could be named line AB or line BA or $\overleftrightarrow{AB}$ or $\overleftrightarrow{BA}$; and line m.

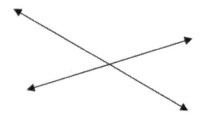

Two lines are considered parallel to each other if, while extending infinitely, they will never intersect (or meet). Parallel lines point in the same direction and are always the same distance apart. Two lines are considered perpendicular if they intersect to form right angles. Right angles are 90°. Typically, a small box is drawn at the intersection point to indicate the right angle.

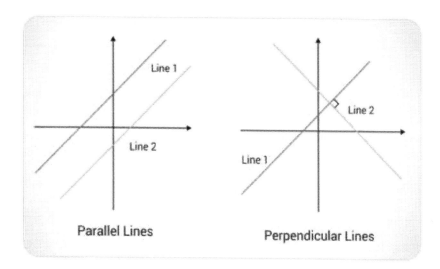

Line 1 is parallel to line 2 in the left image and is written as line 1 || line 2. Line 1 is perpendicular to line 2 in the right image and is written as line 1 ⊥ line 2.

A ray has a specific starting point and extends in one direction without ending. The endpoint of a ray is its starting point. Rays are named using the endpoint first, and any other point on the ray. The following ray can be named ray AB and written $\overrightarrow{AB}$.

An angle can be visualized as a corner. It is defined as the formation of two rays connecting at a vertex that extend indefinitely.

A line segment has specific starting and ending points. A line segment consists of two endpoints and all the points in between. Line segments are named by the two endpoints. The example below is named segment *KL* or segment *LK*, written $\overline{KL}$ or $\overline{LK}$.

Identifying Lines of Symmetry

Symmetry is another concept in geometry. If a two-dimensional shape can be folded along a straight line and the halves line up exactly, the figure is *symmetric*. The line is known as a *line of symmetry*. Circles, squares, and rectangles are examples of symmetric shapes.

Below is an example of a pentagon with a line of symmetry drawn.

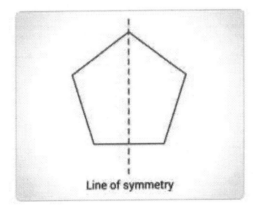

If a line cannot be drawn anywhere on the object to flip the figure onto itself, the object is asymmetrical. An example of a shape with no line of symmetry would be a scalene triangle.

Applying Knowledge of Right Angles

Triangles can be further classified by their sides and angles. A triangle with its largest angle measuring 90° is a right triangle. A triangle with the largest angle less than 90° is an acute triangle. A triangle with the largest angle greater than 90° is an obtuse triangle. The sum of the angles within any triangle is always 180 degrees. Below is an example of a right triangle.

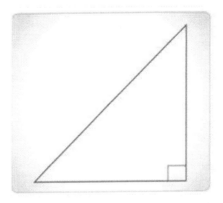

Classifying Two-Dimensional Figures

The vocabulary regarding many two-dimensional shapes is an important part classifying these figures. Many four-sided figures can be identified using properties of angles and lines. A *quadrilateral* is a closed shape with four sides. A *parallelogram* is a specific type of quadrilateral that has two sets of parallel lines that have the same length. A *trapezoid* is a quadrilateral having only one set of parallel sides. A *rectangle* is a parallelogram that has four right angles. A *rhombus* is a parallelogram with two acute angles, two obtuse angles, and four equal sides. The acute angles are of equal measure, and the obtuse angles are of equal measure. Finally, a *square* is a rhombus consisting of four right angles. It is important to note that some of these shapes share common attributes. For instance, all four-sided shapes are quadrilaterals. All squares are rectangles, but not all rectangles are squares.

A triangle consisting of two equal sides and two equal angles is an isosceles triangle. A triangle with three equal sides and three equal angles is an equilateral triangle. A triangle with no equal sides or angles is a scalene triangle.

Polygons can be classified by the number of sides (also equal to the number of angles) they have. Equiangular polygons are polygons in which the measure of every interior angle is the same. The sides of equilateral polygons are always the same length. If a polygon is both equiangular and equilateral, the polygon is defined as a regular polygon.

Angles

Determining Measures of Angles

An angle consists of two rays that have a common endpoint. This common endpoint is called the vertex of the angle. The two rays can be called sides of the angle. The angle below has a vertex at point *B* and the sides consist of ray *BA* and ray *BC*. An angle can be named in three ways:

1. Using the vertex and a point from each side, with the vertex letter in the middle.
2. Using only the vertex. This can only be used if it is the only angle with that vertex.
3. Using a number that is written inside the angle.

The angle below can be written ∠*ABC* (read angle *ABC*), ∠*CBA*, ∠*B*, or ∠1.

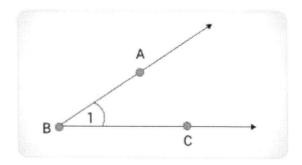

An angle divides a plane, or flat surface, into three parts: the angle itself, the interior (inside) of the angle, and the exterior (outside) of the angle. The figure below shows point *M* on the interior of the angle and point *N* on the exterior of the angle.

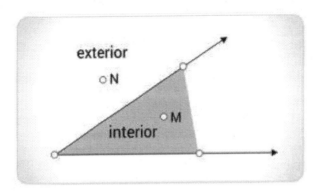

Angles can be measured in units called degrees, with the symbol °. The degree measure of an angle is between 0° and 180°, is a measure of rotation, and can be obtained by using a protractor.

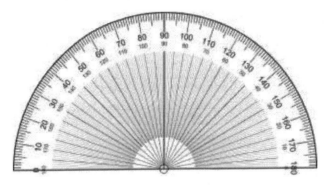

A straight angle (or simply a line) measures exactly 180°. A right angle's sides meet at the vertex to create a square corner. A right angle measures exactly 90° and is typically indicated by a box drawn in the interior of the angle. An acute angle has an interior that is narrower than a right angle. The measure of an acute angle is any value less than 90° and greater than 0°. For example, 89.9°, 47°, 12°, and 1°. An obtuse angle has an interior that is wider than a right angle. The measure of an obtuse angle is any value greater than 90° but less than 180°. For example, 90.1°, 110°, 150°, and 179.9°. Any two angles that sum up to 90 degrees are known as *complementary angles*.

- Acute angles: Less than 90°
- Obtuse angles: Greater than 90°
- Right angles: 90°
- Straight angles: 180°

Determining the Measure of an Unknown Angle

To determine angle measures for adjacent angles, angles that share a common side and vertex, at least one of the angles must be known. Other information that is necessary to determine such measures include that there are 90⁰ in a right angle, and there are 180⁰ in a straight line. Therefore, if two adjacent angles form a right angle, they will add up to 90⁰, and if two adjacent angles form a straight line, they add up to 180⁰.

If the measurement of one of the adjacent angles is known, the other can be found by subtracting the known angle from the total number of degrees.

For example, given the following situation, if angle *a* measures 55°, find the measure of unknown angle *b*:

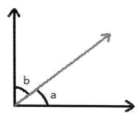

To solve this simply subtract the known angle measure from 90°.

$$90° - 55° = 35°$$

The measure of *b* = 35°.

Given the following situation, if angle 1 measures 45°, find the measure of the unknown angle 2:

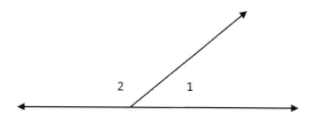

To solve this, simply subtract the known angle measure from 180°.

$$180° - 45° = 135°$$

The measure of angle 2 = 135°.

In the case that more than two angles are given, use the same method of subtracting the known angles from the total measure.

For example, given the following situation, if angle *y* = 40°, and angle *z* = 25°, find unknown angle *x*.

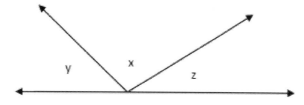

Subtract the known angles from 180°.

$$180° - 40° - 25° = 115°$$

The measure of angle *x* = 115°.

Measurement

Identifying Relative Sizes of Measurement Units

Measurement is how an object's length, width, height, weight, and so on, are quantified. Measurement is related to counting, but it is a more refined process.

The United States customary system and the metric system each consist of distinct units to measure lengths and volume of liquids. The U.S. customary units for length, from smallest to largest, are: inch (in), foot (ft), yard (yd), and mile (mi). The metric units for length, from smallest to largest, are: millimeter (mm), centimeter (cm), decimeter (dm), meter (m), and kilometer (km). The relative size of each unit of length is shown below.

U.S. Customary	Metric	Conversion
12in = 1ft	10mm = 1cm	1in = 254cm
36in = 3ft = 1yd	10cm = 1dm(decimeter)	1m ≈ 3.28ft ≈ 1.09yd
5,280ft = 1,760yd = 1mi	100cm = 10dm = 1m	1mi ≈ 1.6km
	1000m = 1km	

The U.S. customary units for volume of liquids, from smallest to largest, are: fluid ounces (fl oz), cup (c), pint (pt), quart (qt), and gallon (gal). The metric units for volume of liquids, from smallest to largest, are: milliliter (mL), centiliter (cL), deciliter (dL), liter (L), and kiloliter (kL). The relative size of each unit of liquid volume is shown below.

U.S. Customary	Metric	Conversion
8fl oz = 1c	10mL = 1cL	1pt ≈ 0.473L
2c = 1pt	10cL = 1dL	1L ≈ 1.057qt
4c = 2pt = 1qt	1,000mL = 100cL = 10dL = 1L	1gal ≈ 3,785L
4qt = 1gal	1,000L = 1kL	

The U.S. customary system measures weight (how strongly Earth is pulling on an object) in the following units, from least to greatest: ounce (oz), pound (lb), and ton. The metric system measures mass (the quantity of matter within an object) in the following units, from least to greatest: milligram (mg), centigram (cg), gram (g), kilogram (kg), and metric ton (MT).

The relative sizes of each unit of weight and mass are shown below.

U.S. Measures of Weight	Metric Measures of Mass
16oz = 1lb	10mg = 1cg
2,000lb = 1 ton	100cg = 1g
	1,000g = 1kg
	1,000kg = 1MT

Note that weight and mass DO NOT measure the same thing.

Time is measured in the following units, from shortest to longest: second (sec), minute (min), hour (h), day (d), week (wk), month (mo), year (yr), decade, century, millennium. The relative sizes of each unit of time is shown below.

- 60sec = 1min
- 60min = 1h
- 24hr = 1d
- 7d = 1wk
- 52wk = 1yr
- 12mo = 1yr
- 10yr = 1 decade
- 100yrs = 1 century
- 1,000yrs = 1 millennium

Converting Measurements

When working with different systems of measurement, conversion from one unit to another may be necessary. The conversion rate must be known to convert units. One method for converting units is to write and solve a proportion. The arrangement of values in a proportion is extremely important. Suppose that a problem requires converting 20 fluid ounces to cups. To do so, a proportion can be written using the conversion rate of 8fl oz = 1c with x representing the missing value. The proportion can be written in any of the following ways:

$$\frac{1}{8} = \frac{x}{20} \left(\frac{c \text{ for conversion}}{fl \text{ oz for conversion}} = \frac{\text{unknown } c}{fl \text{ oz given}}\right); \frac{8}{1} = \frac{20}{x} \left(\frac{fl \text{ oz for conversion}}{c \text{ for conversion}} = \frac{fl \text{ oz given}}{\text{unknown } c}\right);$$

$$\frac{1}{x} = \frac{8}{20} \left(\frac{c \text{ for conversion}}{\text{unknown } c} = \frac{fl \text{ oz for conversion}}{fl \text{ oz given}}\right); \frac{x}{1} = \frac{20}{8} \left(\frac{\text{unknown } c}{c \text{ for conversion}} = \frac{fl \text{ oz given}}{fl \text{ oz for conversion}}\right)$$

To solve a proportion, the ratios are cross-multiplied and the resulting equation is solved. When cross-multiplying, all four proportions above will produce the same equation: $(8)(x) = (20)(1) \rightarrow 8x = 20$. Dividing by 8 to isolate the variable x, the result is $x = 2.5$. The variable x represented the unknown number of cups. Therefore, the conclusion is that 20 fluid ounces converts (is equal to) 2.5 cups.

Sometimes converting units requires writing and solving more than one proportion. Suppose an exam question asks to determine how many hours are in 2 weeks. Without knowing the conversion rate between hours and weeks, this can be determined knowing the conversion rates between weeks and days, and between days and hours. First, weeks are converted to days, then days are converted to hours. To convert from weeks to days, the following proportion can be written:

$$\frac{7}{1} = \frac{x}{2} \left(\frac{\text{days conversion}}{\text{weeks conversion}} = \frac{\text{days unknown}}{\text{weeks given}}\right)$$

Cross-multiplying produces: $(7)(2) = (x)(1) \rightarrow 14 = x$. Therefore, 2 weeks is equal to 14 days. Next, a proportion is written to convert 14 days to hours:

$$\frac{24}{1} = \frac{x}{14} \left(\frac{\text{conversion hours}}{\text{conversion days}} = \frac{\text{unknown hours}}{\text{given days}}\right)$$

Cross-multiplying produces: $(24)(14) = (x)(1) \rightarrow 336 = x$. Therefore, the answer is that there are 336 hours in 2 weeks.

Solving Problems Concerning Measurements

Problems that involve measurements of length, time, volume, etc. are generally dependent upon understanding how to manipulate between various units of measurement, as well as understanding their equivalencies.

Identifying and utilizing the proper units for the scenario requires knowing how to apply the conversion rates for money, length, volume, and mass. For example, given a scenario that requires subtracting 8 inches from $2\frac{1}{2}$ feet, both values should first be expressed in the same unit (they could be expressed $\frac{2}{3}$ft & $2\frac{1}{2}$ft, or 8in and 30in). The desired unit for the answer may also require converting back to another unit.

Consider the following scenario: A parking area along the river is only wide enough to fit one row of cars and is $\frac{1}{2}$ kilometers long. The average space needed per car is 5 meters. How many cars can be parked along the river? First, all measurements should be converted to similar units: $\frac{1}{2}$km = 500m. The operation(s) needed should be identified. Because the problem asks for the number of cars, the total space should be divided by the space per car. 500 meters divided by 5 meters per car yields a total of 100 cars. Written as an expression, the meters unit cancels and the cars unit is left: $\frac{500m}{5m/car}$ the same as $500m \times \frac{1\ car}{5m}$ yields 100 cars.

For an example manipulating time, Maria is scheduled to take a 90-minute test for her English class. It takes her 25 minutes to get ready and 40 minutes to ride the bus to school. If she begins to get ready at 1:10 p.m., what time will she be finished taking the test?

To find the ending time, all of the elapsed minutes must be totaled and then converted to hours.

$$25 + 40 + 90 = 155 \text{ minutes}$$

The conversion necessary for this problem is that 1 hour = 60 minutes.

The total number of minutes must be converted into hours and minutes, by dividing the total number of minutes by 60.

$$155 \div 60 = 2\ R\ 35$$

The remainder is stated as minutes. So, the total elapsed time is 2 hours and 35 minutes. If Maria begins to get ready at 1:10 p.m., 2 hours from that time is 3:10 p.m., and an additional 35 minutes would add up to 3:45 p.m. Maria can expect to be finished with everything 2 hours and 35 minutes later, at 3:45 p.m.

When measuring length, choosing the right tool to perform the measurement requires determining whether United States customary units or metric units are desired, and having a grasp of the approximate length of each unit and the approximate length of each tool. The measurement can still be performed by trial and error without the knowledge of the approximate size of the tool.

For example, to determine the length of a room in feet, a United States customary unit, various tools can be used for this task. These include a ruler (typically 12 inches/1 foot long), a yardstick (3 feet/1 yard long), or a tape measure displaying feet (typically either 25 feet or 50 feet). Because the length of a room is much larger than the length of a ruler or a yardstick, a tape measure should be used to perform the measurement.

Data Analysis and Personal Financial Literacy

Collecting, Organizing, Displaying, and Interpreting Data

Representing Data

A set of data can be visually displayed in various forms allowing for quick identification of characteristics of the set. Histograms, such as the one shown below, display the number of data points (vertical axis) that fall into given intervals (horizontal axis) across the range of the set. Suppose the histogram below displays IQ scores of students. Each rectangle represents the number of students with scores between a given ten-point span. For example, the furthest bar to the right indicates that two trees are 90 feet tall. Histograms can describe the center, spread, shape, and any unusual characteristics of a data set.

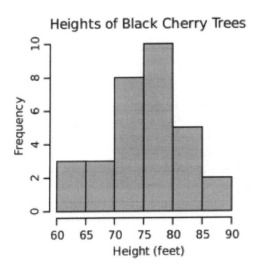

A line plot, also called dot plot, displays the frequency of data (numerical values) on a number line. To construct a line plot, a number line is used that includes all unique data values. It is marked with x's or dots above the value the number of times that the value occurs in the data set.

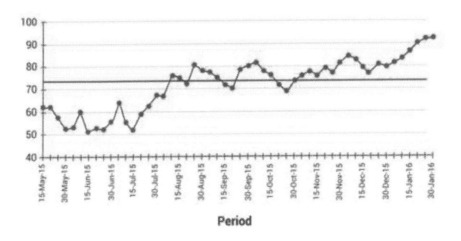

A stem-and-leaf plot is a method of displaying sets of data by organizing the numbers by their stems (usually the tens digit) and the different leaf values (usually the ones digit).

For example, to organize a number of movie critic's ratings, as listed below, a stem and leaf plot could be utilized to display the information in a more condensed manner.

Movie critic scores: 47, 52, 56, 59, 61, 64, 66, 68, 68, 70, 73, 75, 79, 81, 83, 85, 86, 88, 88, 89, 90, 90, 91, 93, 94, 96, 96, 99.

	Movie Ratings
4	7
5	2 6 9
6	1 4 6 8 8
7	0 3 5 9
8	1 3 5 6 8 8 9
9	0 0 1 3 4 6 6 9
Key	6 | 1 represents 61

Looking at this stem and leaf plot, it is easy to ascertain key features of the data set. For example, what is the range of the data in the stem-and-leaf plot?

Using this method, it is easier to visualize the distribution of the scores and answer the question pertaining to the range of scores, which is $99 - 47 = 52$.

Another way to represent data is with a frequency table. These can display the number of times specific answers are given, in order to provide a clearer overall picture of information. The following is the data representing the scores gathered on a patient satisfaction survey, visualized on a frequency dot table.

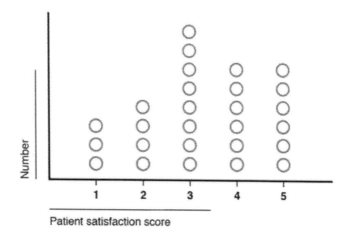

It is clear that the majority of the scores are in the middle, and higher at 2.0 and 2.5.

Solving Problems Using Data

Multi-step problems can be solved using information displayed in a stem-and-leaf plot. For example, the following graph shows the data collected regarding snowfall on top of specific mountains in the Alps.

It can be used to answer the following questions regarding the data.

February		April
	0	2, 4, 9
8	1	
9, 5, 2	2	2, 4, 7, 8
6, 2	3	1, 3
9, 6, 2	4	0, 3, 6, 8, 9
8, 7, 6, 6, 5	5	0
4	6	2

On the left side, 6 | 4 means 4.6 in.   On the right side, 4 | 6 means 4.6 in.

1. Which month had the largest collective snowfall, February or April?
2. How much larger was this snowfall?
3. How many mountains reported more than 4.0 inches of snowfall in February?
4. What is the difference between the lowest reported snowfall in February and the lowest reported snowfall in April?
5. What was the total for the three highest snowfalls in April?

The solution involves adding up the total amount of snowfall in both months individually and finding that February reported more snowfall than April with a total of 64.5 inches. This total was more than the April snowfall by 12.7 inches. There are 9 data points that are higher than 4.0 inches in February. The lowest reported snowfall in February is 1.8 inches, and the lowest reported snowfall in April is 0.2 inches. The difference between the two points is 1.6 inches. The three highest snowfalls in April are 6.2, 5.0, and 4.9. The total of these is 16.1 inches.

Managing Financial Resources

Distinguishing Between Fixed and Variable Expenses

Differentiating between types of expenses in the real world helps to estimate and gauge overall costs for business and living expenses.

A *fixed cost* is an expense that does not change based upon usage or production. For instance, when an individual signs a lease, they agree to pay a certain amount of rent each month. The amount does not change for the time span agreed upon by the renter and the owner. This goes for living space or business space. Many times, businesses pay rent based upon the size of the space they are renting (usually a set price per square foot).

A *variable cost* is one that changes based upon how much of an item is used, or how many products are produced. An example of this is the cost of electricity. The monthly cost of electricity depends upon how much an individual uses, based on a set price per unit. This is also true for a company that produces items that they will sell: the cost of electricity will fluctuate monthly, depending on how much electricity is used to produce goods. For the company, electricity is a *variable expense*.

The difference is that a fixed cost does not depend on any changing item. It is a steady expense, whereas a variable expense depends upon a rate, either on usage or production.

Calculating Profit

In a business situation, there is a production expense and that must be accounted for before calculating any money that will be revenue or profit. *Profit* is the amount of money made after the production cost is accounted for. The supplies necessary to run a business (*operating costs*) are considered an expense. Any money made off of the sale of a business's product after subtracting out the expenses is considered to be a profit.

For example, Matt can make a batch (20 cups) of lemonade out of 4 lemons, which cost Matt $5.00 in total. If Matt earns $1.00 per cup of lemonade, how much profit does he make on a batch of lemonade?

Matt's initial production expenses are $5.00, for the lemons.

Matt's earning from the sale of 20 cups at $1.00 per cup is modeled by the following equation:

$$\$1.00 \times 20 = \$20.00$$

Subtracting the expenses from the earnings will result in Matt's profit:

$$\$20.00 - \$5.00 = \$15.00$$

Matt would have a profit of $15.00 per batch of lemonade.

Angie buys supplies for dog bandanas for $8. Angie can make 5 bandanas with each set of supplies; she sells each bandana for $5. What is Angie's profit per set of supplies?

$$\$5 \times 5 = \$25 - \$8 = \$13$$

Angie earns $13 profit on each set of bandanas she makes.

Describing the Purpose of Financial Institutions

The United States is a capitalist society, where solid banking practices are key. To ensure the security of our banking systems and guarantee that money will retain its value, certain rules and regulations are required to be followed. Banks have to ensure that their lending practices follow fair and impartial guidelines, while also ensuring the quality of investments they make with loans or purchases. Banks are not permitted to make frivolous transactions. Transactions must all be agreed upon by a board that oversees the overall benefit of the investors. Transactions must also be legal and reported through established financial guidelines. Any transactions larger than $10,000 must be reported to the government by banks in order to provide a level of checks and balances that further guarantee money is not being abused or gained through illegal actions.

These rules were put in place to protect investors and to encourage people to invest and trust their money with the banks. All of these rules provide a stronger base for the entire nation. The security offered by the financial banking institutions encourages people to try to save money so as to use their investments at a later date. Setting aside small amounts of money on a regular basis can assist with being able to afford more expensive items such as college or a house. When people save their money within banking institutions, it makes them desirable as a steady stream of investment to the banks. Thus, banks will often be more inclined to provide funding or loans to such people.

Practice Questions

1. It is necessary to line up decimal places within the given numbers before performing which of the following operations?
   a. Multiplication
   b. Division
   c. Subtraction
   d. Fractions

2. Which of the following are units in the metric system?
   a. Inches, feet, miles, pounds
   b. Millimeters, centimeters, meters, pounds
   c. Kilograms, grams, kilometers, meters
   d. Teaspoons, tablespoons, ounces

3. A piggy bank contains 12 dollars' worth of nickels. A nickel weighs 5 grams, and the empty piggy bank weighs 1050 grams. What is the total weight of the full piggy bank?
   a. 1,110 grams
   b. 1,200 grams
   c. 2,250 grams
   d. 2,200 grams

4. A construction company is building a new housing development with the property of each house measuring 30 feet wide. If the length of the street is zoned off at 345 feet, how many houses can be built on the street?
   a. 11
   b. 115
   c. 11.5
   d. 12

5. Which of the following represents one hundred eighty-two million, thirty-six thousand, four hundred twenty-one and three hundred fifty-six thousandths?
   a. 182,036,421.356
   b. 182,036,421.0356
   c. 182,000,036,421.0356
   d. 182,000,036,421.356

6. Which of the following could be used in the classroom to show $\frac{3}{7} < \frac{5}{6}$ is a true statement?
   a. A bar graph
   b. A number line
   c. An area model
   d. Base 10 blocks

7. If you were showing your friend how to round 245,867 to the nearest thousands, which place value would be used to decide whether to round up or round down?
   a. Hundreds
   b. Thousands
   c. Tens
   d. Ten-Thousands

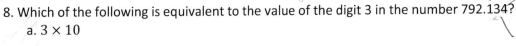

8. Which of the following is equivalent to the value of the digit 3 in the number 792.134?
   a. 3×10
   b. 3×100
   c. $\frac{3}{10}$
   d. $\frac{3}{100}$

9. In the following expression, which operation should be completed first? $5 \times 6 + 4 \div 2 - 1$.
   a. Multiplication
   b. Addition
   c. Division
   d. Subtraction

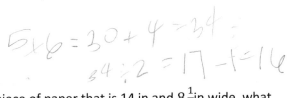

10. In order to calculate the perimeter of a legal sized piece of paper that is 14 in and $8\frac{1}{2}$ in wide, what formula would be used?
    a. $P = 14 + 8\frac{1}{2}$
    b. $P = 14 + 8\frac{1}{2} + 14 + 8\frac{1}{2}$
    c. $P = 14 \times 8\frac{1}{2}$
    d. $P = 14 \times \frac{17}{2}$

11. A grocery store is selling individual bottles of water, and each bottle contains 750 milliliters of water. If 12 bottles are purchased, what conversion will correctly determine how many liters that customer will take home?
    a. 100 milliliters equals 1 liter
    b. 1,000 milliliters equals 1 liter
    c. 1,000 liters equals 1 milliliter
    d. 10 liters equals 1 milliliter

12. An angle measures 54 degrees. In order to correctly determine the measure of its complementary angle, what concept is necessary?
    a. Two complementary angles make up a right angle.
    b. Complementary angles are always acute.
    c. Two complementary angles sum up to 90 degrees.
    d. Complementary angles sum up to 360 degrees.

13. Joshua has collected 12,345 nickels over a span of 8 years. He took them to bank to deposit into his bank account. If the students were asked to determine how much money he deposited, for what mathematical topic would this problem be a good introduction?
   a. Adding decimals
   b. Multiplying decimals
   c. Geometry
   d. The metric system

14. Kassidy drove for 3 hours at a speed of 60 miles per hour. Using the distance formula, $d = r \times t$ ($distance = rate \times time$), how far did Kassidy travel?
   a. 20 miles
   b. 180 miles
   c. 65 miles
   d. 120 miles

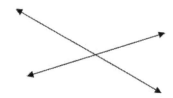

15. Which of the following statements is true about the two lines below?

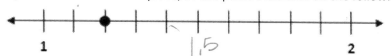

   a. The two lines are parallel but not perpendicular.
   b. The two lines are perpendicular but not parallel.
   c. The two lines are both parallel and perpendicular.
   d. The two lines are neither parallel nor perpendicular.

16. Which of the following numbers is greater than (>) 220,058?
   a. 220,158
   b. 202,058
   c. 220,008
   d. 217,058

17. What is the value, to the nearest tenths place, of the point indicated on the following number line?

   a. 0.2
   b. 1.4
   c. 1.2
   d. 2.2

18. What two fractions add up to $\frac{7}{6}$?

   a. $\frac{2}{3} + \frac{5}{3}$
   b. $\frac{1}{5} + \frac{6}{5}$
   c. $\frac{1}{6} + \frac{6}{6}$ ✓
   d. $\frac{1}{2} + \frac{6}{4}$

19. Which represents the number 0.65 on a number line?

   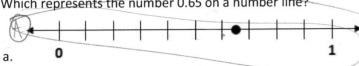
   a.

   b.

   c.

   d.

20. What equation, involving the addition of two fractions, is represented on the following number line?

   a. $\frac{4}{5} + \frac{3}{5} = \frac{7}{5}$ ✓
   b. $\frac{4}{5} + \frac{7}{5} = \frac{7}{5}$
   c. $\frac{3}{5} + \frac{3}{5} = \frac{6}{5}$
   d. $\frac{4}{5} + 1\frac{3}{5} = \frac{7}{5}$

21. What is the product of 26×12?
   a. 78
   b. 202
   c. 302
   d. 312 ✓

22. Which of the following equations is correct?
   a. $123 \div 4 = 33$
   b. $123 \div 4 = 30 + 3$
   c. $123 \div 4 = 30\ R3$ ✓
   d. $123 \div 4 = 3\ R30$

   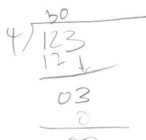

23. What would 307,119 be when rounded to the nearest ten-thousands place?
    a. 307,000
    b. 300,000
    c. 310,000
    d. 320,000

24. If Amanda can eat two times as many mini cupcakes as Marty, what would the missing values be for the following input-output table?

Input (number of cupcakes eaten by Marty)	Output (number of cupcakes eaten by Amanda)
1	2
3	
5	10
7	
9	18

   a. 6, 10
   b. 3, 11
   c. 6, 14
   d. 4, 12

25. Which of the following is not a parallelogram?

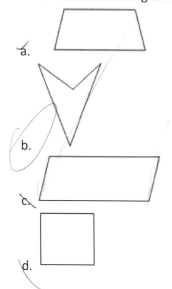

   a.
   b.
   c.
   d.

26. What would the measure of angle 2 be in the diagram below?

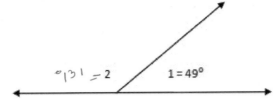

a. 131°
b. 41°
c. 311°
d. 49°

27. Mo needs to buy enough material to cover the walls around the stage for a theater performance. If he needs 79 feet of wall covering, what is the minimum number of yards of material he should purchase if the material is sold only by whole yards?

a. 23 yards
b. 25 yards
c. 26 yards
d. 27 yards

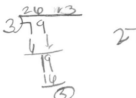

28. The following table shows the temperature readings in Ohio during the month of January. How many more times was the temperature between 36-38 degrees than between 28-32 degrees?

Maximum Temperatures in degrees	Tally marks	Frequency
20 - 22	I	1
22 - 24	JHT II	7
24 - 26	JHT	5
26 - 28	JHT IIII	9
28 - 30	JHT JHT	10

wrong chart

a. 9 times
b. 5 times
c. 4 times
d. 10 times

54

The following stem-and-leaf plot shows plant growth in cm for a group of tomato plants.

Stem	Leaf
2	0 2 3 6 8 8 9
3	2 6 7 7
4	7 9
5	4 6 9

29. What is the range of measurements for the tomato plants' growth?
   a. 29 cm
   b. 37 cm
   c. 39 cm
   d. 59 cm

30. How many plants grew more than 35 cm?
   a. 4 plants
   b. 5 plants
   c. 8 plants
   d. 9 plants

31. Art rents a space in a building downtown for his wicker business. His lease is signed for three years. He pays for materials based upon the number of orders he receives during the month. His electric bill depends upon how many hours he runs the machines in the rented space. Art also pays an assistant an hourly rate, based on the number of orders received. Which of the business costs are fixed, not variable?
   a. Rent
   b. Materials
   c. Electric bill
   d. Assistant's wages

32. It costs Shea $12 to produce 3 necklaces. If he can sell each necklace for $20, how much profit would he make if he sold 60 necklaces?
   a. $240
   b. $360
   c. $960
   d. $1200

55

33. Which is NOT a way that banks can help a capitalist society?
    a. Approving loans
    b. Investing in businesses
    c. Keeping all the money away from people
    d. Keeping money safe and protected

34. What is one of the rules put in place to encourage people to invest their money with a bank?
    a. There are no established financial guidelines that banks must follow.
    b. Transactions larger than $15,000 must be reported to the government.
    c. Transactions must be legal.
    d. Bank managers are responsible for overseeing the quality of investments.

Answer Explanations

1. C: Numbers should be lined up by decimal places before subtraction is performed. This is because subtraction is performed within each place value. The other operations, such as multiplication, division, and exponents (which is a form of multiplication), involve ignoring the decimal places at first and then including them at the end.

2. C: Inches, pounds, and baking measurements, such as tablespoons, are not part of the metric system. Kilograms, grams, kilometers, and meters are part of the metric system.

3. C: A dollar contains 20 nickels. Therefore, if there are 12 dollars' worth of nickels, there are $12 \times 20 = 240$ nickels. Each nickel weighs 5 grams. Therefore, the weight of the nickels is $240 \times 5 = 1,200$ grams. Adding in the weight of the empty piggy bank, the filled bank weighs 2,250 grams.

4. A: 11. To determine the number of houses that can fit on the street, the length of the street is divided by the width of each house: $345 \div 30 = 11.5$. Although the mathematical calculation of 11.5 is correct, this answer is not reasonable. Half of a house cannot be built, so the company will need to either build 11 or 12 houses. Since the width of 12 houses (360 feet) will extend past the length of the street, only 11 houses can be built.

5. A: 182 is in the millions, 36 is in the thousands, 421 is in the hundreds, and 356 is the decimal.

6. B: This inequality can be seen with the use of a number line. $\frac{3}{7}$ is close to $\frac{1}{2}$. $\frac{5}{6}$ is close to 1, but less than 1. Therefore, $\frac{3}{7}$ is less than $\frac{5}{6}$.

7. A: The place value to the right of the thousands place, which would be the hundreds place, is what gets utilized. The value in the thousands place is 5. The number in the place value to its right is greater than 4, so the 5 gets bumped up to 6. Everything to its right turns to a zero, to get 246,000.

8. D: $\frac{3}{100}$. Each digit to the left of the decimal point represents a higher multiple of 10 and each digit to the right of the decimal point represents a quotient of a higher multiple of 10 for the divisor. The first digit to the right of the decimal point is equal to the value $\div 10$. The second digit to the right of the decimal point is equal to the value $\div (10 \times 10)$, or the value $\div 100$.

9. A: Using the order of operations, multiplication and division are computed first from left to right. Multiplication is on the left; therefore, the multiplication should be performed first.

10. B: Perimeter of a rectangle is the sum of all four sides. Therefore, the answer is $P = 14 + 8\frac{1}{2} + 14 + 8\frac{1}{2} = 14 + 14 + 8 + \frac{1}{2} + 8 + \frac{1}{2} = 45$ square inches.

11. B: $12 \times 750 = 9,000$. Therefore, there are 9,000 milliliters of water, which must be converted to liters. 1,000 milliliters equals 1 liter; therefore, 9 liters of water are purchased.

12. C: The measure of two complementary angles sums up to 90 degrees. $90 - 54 = 36$. Therefore, the complementary angle is 36 degrees.

13. B: Each nickel is worth $0.05. Therefore, Joshua deposited $12{,}345 \times \$0.05 = \$617.25$. Working with change is a great way to teach decimals to children, so this problem would be a good introduction to multiplying decimals.

14. B: 180 miles. The rate, 60 miles per hour, and time, 3 hours, are given for the scenario. To determine the distance traveled, the given values for the rate (r) and time (t) are substituted into the distance formula and evaluated: $d = r \times t \to d = (60mi/h) \times (3h) \to d = 180mi$.

15. D: The two lines are neither parallel nor perpendicular. Parallel lines will never intersect or meet. Therefore, the lines are not parallel. Perpendicular lines intersect to form a right angle (90°). Although the lines intersect, they do not form a right angle, which is usually indicated with a box at the intersection point. Therefore, the lines are not perpendicular.

16. A: This choice can be determined by comparing the place values, beginning with that which is the farthest left; hundred-thousands, then ten-thousands, then thousands, then hundreds. It is in the hundreds place that Choice A is larger. Choice B is smaller in the ten-thousands place, Choice C is smaller in the tens place, and Choice D is smaller in the ten-thousands place.

17. C: The number line is divided into 10 sections, so each portion represents 0.1. Because the number line begins at 1 and ends at 2, the number in question would be between those two numbers. Since there are only two portions out of ten marked, this represents the number 1.2. All other choices are incorrect due to a misreading of the number line.

18. C: To add fractions, the denominator must be the same. This is the only choice with both denominators of 6. Adding the numerators totals 7, for a fraction of $\frac{7}{6}$. Choice A equals $\frac{7}{3}$, Choice B equals $\frac{7}{5}$, and Choice D equals $\frac{8}{4}$ or 2.

19. A: The number line is divided into ten portions, so each mark represents 0.1. Halfway between the 6th and 7th marks would be 0.65. Choice B shows 0.55, Choice C shows 0.25, and Choice D shows 0.45.

20. A: The light gray portion represents $\frac{4}{5}$ and the dark gray portion represents $\frac{3}{5}$, to total $\frac{7}{5}$. Choice B is not correct because it misrepresents the dark gray portion. Choice C is not correct because it misrepresents the light gray portion. Choice D is not correct because it includes the 1 with the dark gray portion.

21. D: This answer is the only one that carries out proper multiplication to get the correct result (as seen below).

```
    2¹6
   ×1 2
    5 2
  +2¹6 0
   3 1 2
```

The calculation for Choice A is incorrect, as it does not place the zero marker for the second row of numbers being multiplied. The calculation for Choice B is incorrect, as it does not carry any of the ones necessary to multiply the numbers. The calculation for Choice C is incorrect, as it does not carry the one in the multiplication portion of the problem.

22. C: If an array were to be used, 123 items could be divided up into 4 groups of 30, with 3 left over. Choices A, C, and D are misrepresentations of the correct grouping and not equal to 30 with a remainder of 3.

23. C: In order to round to the nearest ten-thousands place, the number to the right of that digit (7) must be observed. Since 7 is greater than 4, the ten-thousands place number would round up, therefore 0 becomes 1. Choice A rounds to the thousands place, Choice B incorrectly rounds to the ten-thousands place, and Choice D also incorrectly rounds to the ten-thousands place.

24. C: The situation can be described by the equation $? \times 2$. Filling in for the missing numbers would result in $3 \times 2 = 6$ and $7 \times 2 = 14$. Therefore, the missing numbers are 6 and 14. The other choices are miscalculations or misidentification of the pattern formed by the table.

25. B: This is the only shape that has no parallel sides, and therefore cannot be a parallelogram. Choice A has one set of parallel sides and Choices C and D have two sets of parallel sides.

26. A: The way to calculate the measure of angle 2 is to subtract angle 1 from the measure of a straight line (180°). $180° - 49° = 131°$. Choice B subtracts the value of angle 1 from 90°, Choice C subtracts the value of angle 1 from 360°, and Choice D mistakenly labels angle 2 as equal to angle 1.

27. D: In order to solve this problem, the number of feet in a yard must be established. There are 3 feet in every yard. The equation to calculate the minimum number of yards is $79 \div 3 = 26\frac{1}{3}$.

If the material is sold only by whole yards, then Mo would need to round up to the next whole yard in order to cover the extra $\frac{1}{3}$ yard. Therefore, the answer is 27 yards. None of the other choices meets the minimum whole yard requirement.

28. C: To calculate this, the following equation is used: $10 - (5 + 1) = 4$. The number of times the temperature was between 36-38 degrees was 10. Finding the total number of times the temperature was between 28-32 degrees requires totaling the categories of 28-30 degrees and 30-32 degrees, which is $5 + 1 = 6$. This total is then subtracted from the other category in order to find the difference. Choice A only subtracts the 28-30 degrees from the 36-38 degrees category. Choice B only subtracts the 30-32 degrees category from the 36-38 degrees category. Choice D is simply the number from the 36-38 degrees category.

29. C: The range of the entire stem-and-leaf plot is found by subtracting the lowest value from the highest value, as follows: $59 - 20 = 39$ cm. All other choices are miscalculations read from the chart.

30. C: To calculate the total greater than 35, the number of measurements above 35 must be totaled; 36, 37, 37, 47, 49, 54, 56, 59 = 8 measurements. Choice A is the number of measurements in the 3 categories, Choice B is the number in the 4 and 5 categories, and Choice D is the number in the 3, 4, and 5 categories.

31. A: His rent is a fixed rate, paid over the course of his three-year lease. All other expenses depend upon the number of orders, the use of services, and the hours necessary to complete the work; these are all called variable expenses.

32. C: In order to calculate the profit, an equation modeling the total income less the cost of the materials needs to be formulated. $60 \times 20 = \$1,200$ total income. $60 \div 3 = 20$ sets of materials. $20 \times \$12 = \$240$ cost of materials. $\$1,200 - \$240 = \$960$ profit. Choice *A* is not correct, as it is only the cost of materials. Choice *B* is not correct, as it is a miscalculation. Choice *D* is not correct, as it is the total income from the sale of the necklaces.

33. C: Banks approve loans to people and businesses, and banks also keep money safe and guarantee it will be there when people who have accounts want to make withdrawals. Banks do not keep all the money locked up or away from people because the movement of money in and out of banks adds to the health of a capitalist society.

34. C: Transactions must be legal and reported to the government using established financial guidelines, which include reporting to the government all transactions over $10,000. A board oversees the quality of investments and the fairness of transactions.

Photo Credits
The following photo is licensed under CC BY 2.5 (creativecommons.org/licenses/by/2.5/)

"Black cherry tree histogram" by Mwtoews
(https://commons.wikimedia.org/wiki/Histogram#/media/File:Black_cherry_tree_histogram.svg)

FREE Test Taking Tips DVD Offer

To help us better serve you, we have developed a Test Taking Tips DVD that we would like to give you for FREE. **This DVD covers world-class test taking tips that you can use to be even more successful when you are taking your test.**

All that we ask is that you email us your feedback about your study guide. Please let us know what you thought about it – whether that is good, bad or indifferent.

To get your **FREE Test Taking Tips DVD**, email freedvd@studyguideteam.com with "FREE DVD" in the subject line and the following information in the body of the email:

    a. The title of your study guide.

    b. Your product rating on a scale of 1-5, with 5 being the highest rating.

    c. Your feedback about the study guide. What did you think of it?

    d. Your full name and shipping address to send your free DVD.

If you have any questions or concerns, please don't hesitate to contact us at freedvd@studyguideteam.com.

Thanks again!

Made in the USA
Middletown, DE
12 April 2018